How Did I Miss That????

SUPPORT FOR PARENTS WHO HAVE LOST
A CHILD TO SUICIDE

Cindy Harris

Battle Press
SATELLITE BEACH, FLORIDA

How Did I Miss That????

Support For Parents That Have Lost A Child To Suicide

Copyright © 2021 by Cindy Harris.

All rights reserved. No part of this book may be used or reproduced by any means, graphic, electronic or mechanical, including photocopying, recording, taping or by any information storage retrieval system without the written permission of the author or publisher except in the case of brief quotations embodied in critical articles and reviews.

Battle Press books may be ordered through booksellers or by contacting:

Battle Press
steve@battlepress.media
www.battlepress.media

ISBN: 978-1-7378-2380-3 (softcover)
ISBN: 978-1-5136-8541-0 (eBook)

Library of Congress Control Number: 2021919328

First Edition

Disclaimer

This book is written from my own experience of the loss of my only child, Victoria. Everyone's experience may be different. I wrote this book so that those suffering through this trauma may find these suggestions helpful. The journey for survivors of suicide victims is a long and often brutal one that affects one's life forever. If this book helps one person then I have done well. The suggestions I offer here are purely my own. My heart and empathy go out to all that have had to travel this path.

Dedication

To all the wonderful people who helped me through the worst days of my life and continue to help me. My mom and dad, Grace, Cappy, and Joe. These five were by my side in the early days of my loss. Without them, I don't know where I would be. My sister, Linda, who kept me on my path to spirituality. Christina, who has lent an empathetic ear. Joseph, Ron, Barbara, Kristi, Skye, Mike and Val for their endless support. My two mediums, Tina Vesely, who proved to me Vicki was still with me and walks with me daily, watching and listening to me. Shane Gadd, my holistic medium, who has helped me heal from the raw trauma with endless encouragement with his beautiful talent of passing on messages from spirit. My good friend, Frances Hynes, having leaned on each other, suffering the same loss, picking the other up at our lowest moments. Knowing intimately each other's pain, having experienced the loss of our children, Vicki and Oisin. To my precious Vicki, for all you taught me in the short time you were here, I will always love you. I will see you again on the other side.

To the memory of my mother, Faith, who has gone to be with Vicki, please kiss her and watch over Vicki until I can be with you both. I love and miss you both so very much.

How Did I Miss That????

Table of Contents

Chapter 1 Vicki's Arrival 7
Chapter 2 Moving to the Suburbs 12
Chapter 3 The Day My Life Would Change Forever 19
Chapter 4 Accepting the Unacceptable 25
Chapter 5 Vicki's Legacy 29
Chapter 6 Leaning on Spirituality to Survive 33
Chapter 7 Resolution 36
Chapter 8 Helpful Suggestions in an Unimaginable Situation 38
Chapter 9 Outlook on an Uncertain Future 43
Chapter 10 Thoughts Start to Change - Need a Retreat 46
Chapter 11 Regrets 51
Chapter 12 Forgiveness. Is It Really Possible? 56
Chapter 13 Looking Back over Vicki's Twenty-Six Years. 59
Chapter 14 Triggers and Tailspins 61
Chapter 15 Afterthoughts 64
Chapter 16 Wishes 67
Chapter 17 Visits from Vicki Through the Dream State ... 70
Chapter 18 How to Support someone who has Lost someone to Suicide 76

Chapter 1
Vicki's Arrival

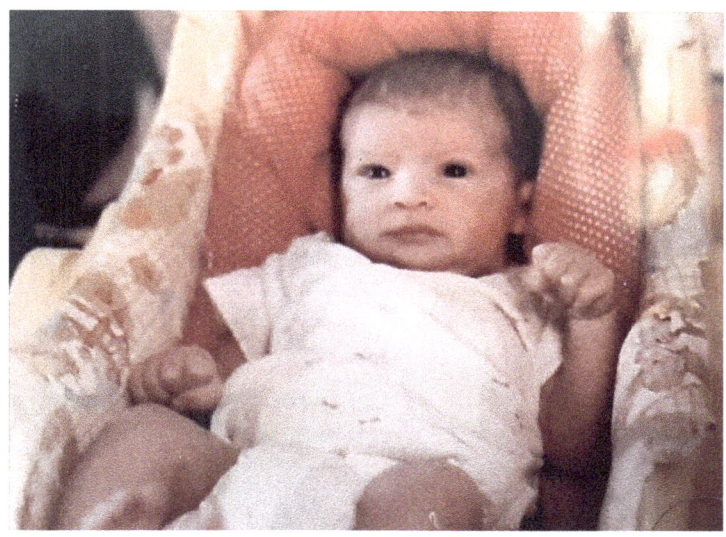

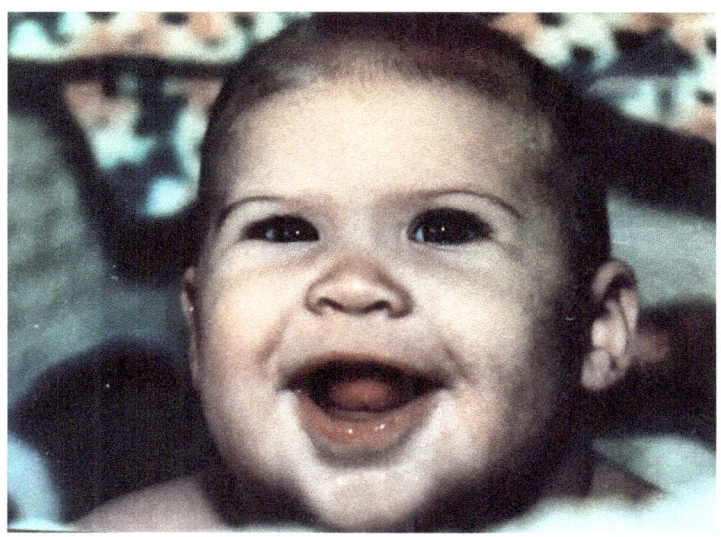

How Did I Miss That????

When I met Vicki's father, I was twenty-one years old and divorced once already. My first husband and I were married at nineteen and really, we were just too young. We weren't even married a year when it came to an end. I met my second husband, still very young. Thinking I could change the world, fearless and able to work through anything! We lived together for a year before getting married. I think the first ten years were okay, but from the start, cracks were showing. The marriage had difficulties. My husband, I believe, suffered from Post-Traumatic Stress Disorder from the Vietnam War, among other difficulties, including anger issues and alcohol addiction. I thought if I loved him enough, he wouldn't be so angry, but there wasn't enough love in the world to fix it all. I married him anyway.

We began looking into buying a house in New Windsor, New York, and found a beautiful ranch home with three bedrooms that we bought in 1987. Fast forward almost two years, our daughter Victoria Lauren Harris was born on April 9th, 1989. Six pounds, six ounces and beautiful. The pregnancy was a difficult one for me. There was a lot of pain and I had experienced a very bad allergic reaction to the HCG hormone that the body develops during pregnancy. I was extremely itchy and scratched the skin on my legs, arms, and back until it was raw. There was a lot of stress too. My job was very demanding and my husband had gotten himself fired, so there were problems with insurance coverage. I was always tired and waiting for the other shoe to drop. It was a constant emotional rollercoaster; I was always striving for unattainable perfection. When Vicki was born, I suffered from post-partum depression horribly, which I never heard of or expected. I cried a lot and was not even sure as to why. No one ever mentioned postpartum symptoms to me; not in the childbirth classes, the hospital, or by family members. I had never ever heard of this. On the second day home from the hospital, my husband declared if I did not stop crying, he was going to go live in the basement, and with that statement, so began the start of a lot of unhappiness. When Vicki was young, my husband seemed to take well to having a baby around. He showed her love and affection, changing diapers, and took turns with hourly feeding. We equally shared this as we began raising our beautiful daughter. As she grew, things became more challenging. More patience is required when you have a two-year-old, hence the term terrible-twos. My husband found a job and decided I should stay home with Vicki for her first two years. Instead of paying for childcare, he thought it would be better for her and easier for us. I was okay with this and it was generous for him to take on that responsibility, so I thought. Thinking back, it really was the best two years of my life. However, the mortgage was increasing and I needed to get back to work. Vicki was growing quick and was growing out of her clothes quickly. She needed many things, as all babies do. I found a job at a local supermarket and began to contribute to the expenses of the household again. I still had some cash set aside which I had planned for later in life, retirement for us, college for Vicki, those sorts of things. However, later I found that plan would take a completely different path. When I went back to work, I began to make money, sometimes more than my husband. I even occasionally worked overnights to make extra. This was not so great for my already difficult marriage.

Vicki was an adorable little girl; pure sparkle and sunshine. Her laugh was contagious and she had a huge sense of humor. Between the very young ages of three and four she became so very independent. It was shocking the way she caught onto things so quickly. She was so smart. She loved cartoons, dancing with me, having me chase her, swimming in the pool, and catching lightening bugs with her dad. My love for her was huge and grew everyday more so. Vicki was also a little escape artist, especially when getting her diaper changed! She would run down our long hallway naked, screaming, and giggling with delight knowing I was hot on her tail. As she learned to talk, words didn't always sound right. Vicki loved her apple juice, but she couldn't say juice and would instead say "deuce". Funny, the things you remember. I loved dressing her in pretty dresses. She would quickly run and grab a pair of my shoes from my closet and go clunking around the house with them.

How Did I Miss That????

Vicki seemed to love being outside the most. The backyard was her place, but the park was even better. Treasures there were limitless. Her favorite holiday was Halloween. We would trick-or-treat and would walk until my feet hurt! When it rained, we would be indoors. There was lots to do there too. She would sit on my foot and go for a ride as I shuffled through dinner, laundry, cleaning, and washing dishes. Sometimes she would just watch one of her favorite cartoons and dance to the music, rocking back and forth. Other times she would pretend to be a mad scientist doing all kinds of experiments in the kitchen sink with spices and food colorings. These are some of my most cherished memories.

Chapter 2
Moving to the Suburbs

My husband and I lived in Rockland County before we bought our home in New Windsor. It was more country-like and a bit boring but I did enjoy owning my own home, and thought it would be a good place to raise a child. Vicki would need to have her own room. We all needed to learn to adjust to new surroundings. There were many more adjustments to be made as Vicki grew. Vicki had difficulty with her vision, so I needed to have that checked. I took her to my Uncle Bill who worked with an Optometrist and as suspected, Vicki needed glasses. She was so upset. She said she felt ugly wearing them. They would be necessary for school and so I would have to remind her every day to bring them. The first day of school arrived, we laid out the clothes she wanted to wear to be ready the night before. Vicki was excited to begin school as making friends was something she looked forward to. Vicki had gone to preschool the year before, but the class was small and most of the children already knew each other so she found it difficult to fit in. The first day of school finally came and off we went to Temple Hill Elementary School. I walked up the sidewalk with her, suddenly she broke away and went running as she waved goodbye to me. I quickly ran to grab her as she didn't even know where her classroom was. As I left that day, walking down the sidewalk, tears stained my cheeks as I was witnessing my little girl growing up. Vicki adapted to the daily routine of school, but never seemed to fit in with most of the other kids. She was quirky and I adored that but it did create a hardship for her. She struggled to make friendships and I figured the next best option was to consider adopting a pet.

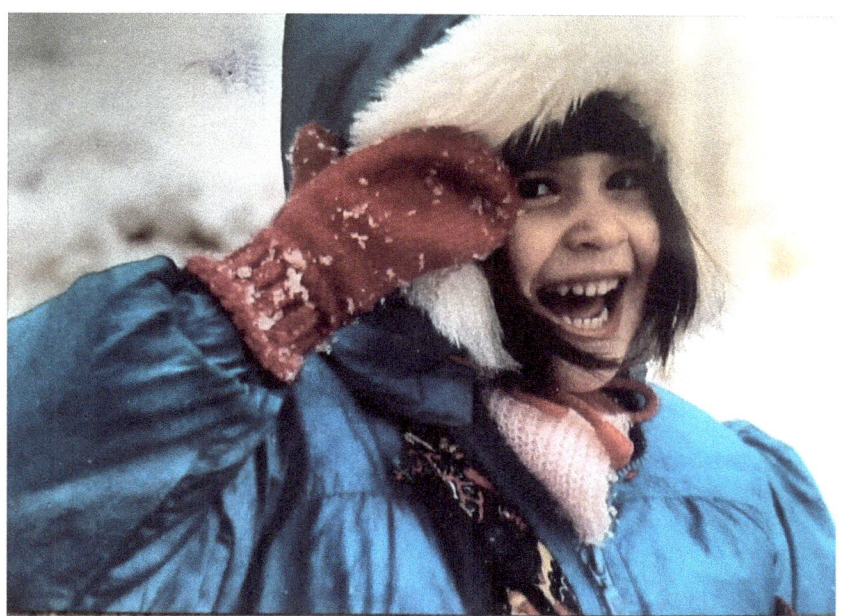

I had an iguana named Iggy when Vicki was a baby, but he had passed away at the age of fourteen. It was some time before Vicki decided to ask the question about adopting a pet. Maybe a cat and so, we did. I took her to a woman who had many rescues and I let her choose one. Vicki picked an older, very sweet and patient cat. She named her "Princess Kitty Jingles". Vicki loved Princess. When Princess passed, Vicki was still very young. I knew Princess was dying, as her health was deteriorating in her old age. Princess was at least fifteen to twenty years old according to the vet. Her teeth were not in good condition and the shelter is where she spent most of her life. The day she passed away the sun came through the front room window very brightly. I laid her in the sun, and took Vicki to school. I told her I would look after her and I did. Princess took some water, but she wouldn't eat. When I brought Vicki home from school, I told her to go sit with Princess because she'd be going to heaven soon. Within a few minutes she came to me with tears streaming down her face and said "Princess is dead, mommy." I asked her how she knew. She said "I saw her take her last breath." So, I said, come on, let me check as we headed for the front room. I really couldn't believe it happened that way. It was as if Princess had waited to say goodbye to Vicki. I didn't realize how much this affected Vicki's outlook on life until I read in a journal she had left in her room. She wrote about her sadness after that day. I then realized the depth that traumatic day had on her life. After reading that entry, I cried, myself. She thought we made fun of her for her choice of such a lengthy name for Princess. She expressed her undying love for Princess and what life would be like without her. Our amusement over her choice of name hurt her, now it hurt me. I was so sorry for ever laughing. Now there was nothing I could do. Words cannot be taken back once said. I will regret that forever. I wish I had gently explained things better to her about death and heaven, as well. Young children do understand more than we realize, but their perception is undeveloped. I can only say I was young and learning, myself. That is really my only excuse.

How Did I Miss That????

As Vicki approached adolescents, she met her first long-term friend, Christina. She was a big part of Vicki's life and the two of them remained inseparable for quite some time. They often took solace in one another's company and confided with each other on many things. There are times, as I would learn, moms are not always wanted around as their child becomes older. I began to notice a change in Vicki after the age of twelve. Looking back, I should have questioned that more. Vicki just seemed to be always tired, even after a whole nights rest. Her eyes seemed darker even after being well-rested. She seemed somewhat depressed, but not always. Preteen girls always seem to be moody and I figured as she grew that would change. I was wrong. I should have looked closer and I regret not having done so. Having been a young girl, myself, I understood that hormones could have a negative effect on one's self-esteem. I thought as she got older, she would adapt to it as most young ladies do, but it was to become much worse. In hindsight, it may have been better to have had her evaluated by a doctor for depression. Had I realized the magnitude of things coming I certainly would have. Once kids turn eighteen though, your rights as a parent change. The thing is, I have met many parents who did have their children seen by doctors, and the child's choice of suicide still happened. Medication and therapy, although helpful can be preventative, but no guarantee. But what is a guarantee, is that you will always second guess yourself and every choice about your child's life that you made until you pass yourself. Judging right or wrong about anything now, I am really unable to do. There really is no right or wrong, only reactions to actions. I have become much more empathetic. Sometimes you don't know what a person is going through. Some people can hide things very well. This was my daughter.

It was the year 2002 when I finally decided some change needed to happen. Vicki was always sad; I was always sad. My husband and I just couldn't get along anymore. All the arguing and fighting had to stop. I told him it was time to make things better for myself and Vicki and I wanted a divorce. In 2003 it became final. Vicki was about to enter high school so we moved in with a friend and co-worker of mine until I could get things sorted. By the following June, the child support started to come late and short. My husband and I continued to argue about that. I needed and used every bit of that child support to buy and pay for all the things a young girl needed, but soon his help would stop altogether when he once again lost his job and didn't work for quite some time. Vicki was not even out of high school when her father moved to another state, many miles away with his new wife. He left without a word. I remember, she went to his home with Christmas gifts in her arms and returned home within a few minutes. Hurt and crying, gifts still in her hands and saying she was going to return all of them. She then told me "Dad moved and couldn't even be bothered to tell me!" Her rage was over the top. She was so angry and hurt. I will never forget the pain in her eyes. I was so angry that he hurt her that way. There were no words. I couldn't even console her; she would have no part of that. My resentment towards him grew.

Vicki still had bouts of depression that were pretty intense. She always seemed to bounce back though. I never thought she would harm herself. I gave her space and privacy to work things out. It seemed to be what she really wanted. I thought she would come to terms with her emotions as I had when I was younger, but she was just learning to hide it better. I stayed in her ear even if she didn't want me there. I wanted her to know I was there for her, but she always pushed me away. I wasn't enough in the end, unfortunately. Vicki didn't like being told what to do and being told "no" even more. She also didn't like having to apologize, but she did like to make people laugh. She was intelligent and very pretty. She had a

How Did I Miss That????

quick temper and a lot of sass. Her friends were the biggest and most important part of her life, and she had many as she grew older. She was always doing her best to make them laugh while hiding her own pain inside. Her deep sadness was turned internally so no one could view it. I had no idea how deep and dark the sadness could get for her. I would find out just how bad things were in the near future. Vicki grew tall, a long-legged beauty. She brought me many proud moments. Graduations, first steps, good grades, personal growth, an eye for detail and a unique sense of fashion. She was so fearless, which frankly scared the crap out of me. She loved being outdoors and was very adventurous. She was an avid reader, with a snarky sense of humor and often stood out in a crowd with bold makeup and hair colors.

By the year 2008 Vicki had entered college. She did very well and made good grades. New friends and faces made a big difference, but she had changed her degree twice, from forensics to computer securities and felt she needed a fourth year of schooling. She wanted an Associate's Degree and to further her education in the field of law. Her depression seemed to increase after graduating college and the expenses of student loans and bills began to weigh heavily on her. Vicki found it difficult to save money because she found temporary comfort in material possessions.

Vicki also had difficulties with romantic relationships. She wanted so badly to find a forever-partner even at her young age but none seemed to last. I think her main focus was to be in a relationship so that she didn't have to focus on the loneliness and pain she often felt. When she couldn't achieve a long-lasting relationship, it really affected her depression and self-esteem. I wish Vicki could have learned to love herself. Her father's and my love, or lack thereof, was not what she deserved in life and it certainly wasn't a good representation of a healthy relationship. She had a serious boyfriend that she, I believe, genuinely loved near the end of her life, but it did not last. I believe that it was the end of this relationship accompanied with her struggles with mental health that extinguished her flame. I believe that this is the relationship she hoped would move her forward in life and bring her all she thought she would ever want. Vicki also hid this relationship from me. It is still something I struggle with. I very much wish she would've talked with me about how hurt she was but Vicki hid everything. And so, she did what she always did which was to suffer silently.

Chapter 3

The Day My Life Would Change Forever

May 31, 2015 started for me at three in the morning for a trip planned months before. There was a bus to catch to the Dover Race Track which was a day at the races and to return at midnight. It's a long day but usually I can catch a nap on the bus. I opened Vicki's door and called up the stairs to her before I left. Her room was on the second floor of my home. I asked if she could hear me but in my head I thought of course she can! It's 3 am and I'm shouting up the stairs. How could she not? I said "Please Vicki, don't forget to feed the cat and bring the mail in, okay?" I then said "See you tonight, have a good day." I heard a groggy voice reply "Okay, bye." I closed her bedroom door and rushed off to the car to catch the bus. That would be the last time I would hear my daughter's voice.

We all arrived at the race track and had a bite to eat at the barbeque they provided. It was a beautiful day being that it wasn't too hot or cold. As we found our seats in the stadium and settled in, I checked my phone for text messages, but there wasn't any. Vicki was a late sleeper, and she worked the night shift. I wouldn't call and wake her. I was in the habit of checking for her texts. The races are very loud, making talking on the phone nearly impossible. After the race we all headed back to the bus. As the people got on, they took a head count and found one person was missing. He had wandered off with some people and decided to stay with them. We wound up leaving over an hour late without him. Most of the people were becoming impatient with waiting and it was decided by the coordinator of the trip that he would return the following day to pick him up. So, we were on our way. I received and sent a few messages to and from Vicki on the way home. It began to rain very hard that night. I let Vicki know approximately what time we'd be back and when I'd be home.

I arrived in my driveway at 11:40 pm. As I pulled in, I was so relieved to see her car in the driveway. I remember saying out loud, "Oh good, at least I don't need to worry when Vicki will be home and I can finally rest." I dropped my bags and cooler on the front porch and rushed to the bathroom. Before I even finished, there was a hard and heavy knock at the door. I couldn't believe someone was knocking on my door that hard so close to midnight. I opened the door to two police officers who said they were there to do a welfare check on Vicki. I was bewildered. Confused, I told them "Sure, she's up in her room." They ran up the stairs, but within seconds they came back down.

They told me that she wasn't up there. I asked if they had checked her bathroom. I said "She must be up there; her car is in the driveway." At that moment, they turned and bolted out the front door. It was pouring rain as I followed right behind them. I was so confused. I was so tired from the long trip we had just come back from, but adrenaline was starting to fuel me. I never expected what would happen next. As they opened the doors to her car there she was. A horrible rotten egg smell filled the air. I couldn't believe what was happening before my eyes. I thought "My god, no! My precious baby, my only child, my only daughter. This can't be happening!" They pulled her out of the car onto the driveway. Some of the poor police officers began to become ill. I patted her face, still warm to the touch. I started to scream, "Wake up, Vicki! Come on honey, wake up!" The rain was furiously coming down on us. The motion detector on the side of the house kept going on and off. The officer I was with asked me if there was a way to keep it on but I didn't know how. I asked him what that smell was and he said he didn't know, but asked me to help him move her away from the car. Together we pulled her a bit further over and her head dropped back to the ground, her eyes opening slightly. I thought, "Thank god, she's going to be okay, she has to be." They raced to try to use paddles on her chest but the rain was making that difficult. I kept grabbing at her face, hands, arms, and body, begging her to talk to me. She was still warm and soft. I needed her to say something or just let me know she could hear me. Another police officer came and pulled me away and said I needed to move back so they could help her. He placed me in the police car and closed the door. After a minute I realized I couldn't open the doors. I watched through the back window of the car, the horror unfolding as the rain drops distorted my view. I watched as they put her in an ambulance. I tried to call my mother and father many times while I sat in that police car. My hands shook so hard I couldn't hit the numbers on my phone. I finally reached them and hollered into the phone, crying, "Something's happened to Vicki, please go to Saint Luke's Hospital!" They said they would meet me there. I sat as shock completely overtook me. It felt like an eternity had passed since that door on the car closed. I started to bang on the window when the officer finally came back to the car. The flashing of all the lights on the many police and emergency vehicles are burned into my memory. I asked the officer as we sped away if she was okay but he said he didn't know. I asked, "How could you not know? You just came from the ambulance!" He just repeated he didn't know.

We arrived at the hospital in five minutes. My mom and dad were already there. I was numb, not crying yet, but in shock from what I had just witnessed. Not understanding, and still believing all would be okay. We were at the hospital; they would surely be able to take care of her. She was only twenty-six. We waited in a room they had us sit in for what seemed like forever. Then the doctor came. I could tell by the look on her face that my life was about to change forever. In that moment, she said "Ms. Harris, I've been doing this a long time, but your daughter left me nothing to work with." I responded, "What are you talking about? I don't understand what you are saying? What do you mean?" I heard her words, but I was bewildered, not believing my ears. Tears started forming now. She said "I'm sorry." "What?" I replied. I wanted to see her with my own eyes, but she said because of the way she ended her life that it wouldn't be safe. I demanded, I didn't care, I wanted to see her but they would not bring me to her. Sometime after that I made a call to her father telling him what I had just heard and what happened prior. I have no recollection though of what I actually said. I only remember making the call. My parents and I finally left the hospital. All of us in disbelief and shock, not really believing what just happened. As I stepped off the curb and into the parking lot, the rain had stopped. I looked up to the sky and felt my mind crack in half. I dropped to the ground and howled in pain. My world changed forever and my tears continue to flow to this day. I've cried an ocean. My dad stayed with me that night but I didn't sleep. I had been awake for over twenty-four hours, since we left for the race track the day prior. I couldn't eat, feeling a sickness in my stomach. I couldn't drink or speak. All I could do was cry, out loud and hard. I was broken. Within a few days I would bury my only child. I began to beat myself up with blame. How could I have missed this? Should've, could've would've filled my mind. I was unable to think clearly and blame overwhelmed me. My thoughts were racing a million different directions. My body was empty, I couldn't process anything. There was a gaping hole in my chest where my heart once was. It hurt as if someone had poured rock salt in it. With the help of an angel named Grace, a funeral was planned. There are many details I simply don't remember. I think that's the body's way of protecting itself. I went through the wake and funeral on auto-pilot. I spoke to each of the many people who attended her funeral. At times I felt I would black out, but I had to keep going. Her wake was a blur. There were so many people who came and I sent her off in the most beautiful way I could. She was covered in white roses, white challah lilies, and purple orchids. Vicki loved the color purple. Her father was there with his wife, but not much was said between us. He did hug me at the end and told me to go to the doctor to get some medication because he said I would need it. He then left.

Vicki was my reason for doing everything I did. I worked hard to give her more than I ever had. I tried hard to make her happy but I felt that I had failed. I was always her mom, though, so I sent her off in blankets of flowers, I kissed her goodbye and continue to pray she found the peace she so desperately needed. A huge part of me wanted to crawl in that casket with her. I thanked God for Grace's help and asked him to hold Vicki in his arms for me. I was a zombie and felt as if I could not function, but with the support of family and friends, I somehow pulled through it. Currently, my life has changed completely from the life I once knew. Unrecognizable, really. I am not the same person I once was. I thought the pain would've surely killed me by now, But I still continue to wake each morning. My life has continued, just not in the way I thought it would. Now, my purpose in life seemed uncertain and empty. I now believe, after much thought, I've been placed here on Earth to help people and develop my spirituality. I try not to be angry about not being able to experience the things I had rightfully expected to experience, and never will. It becomes difficult when having to listen to others tell stories

How Did I Miss That????

in detail about the joy and happiness they get from their children and grandchildren. While I understand, it can be very difficult at times. A thin layer of skin covers the hole where my heart once was, but it can unexpectedly open without warning with every day that passes.

Chapter 4
Accepting the Unacceptable

How is one supposed to accept this?! I can't, I won't! Broken, raw, refusing to understand anything that has happened, not being able to change it. For if I could, it would never had happened at all. I couldn't understand what people were saying to me. Not sleeping or eating probably didn't help. Sleep was impossible. I was still waiting for her to come home and I had no desire to eat. I just couldn't. I had difficulty even breathing at times. Numb and crying, waiting to wake up from this horrible nightmare. The first two weeks after I didn't even get out of bed except to shower and use the toilet. Weak, sick, and in pain, trying to process what had happened to my daughter. My throat was tight and speaking was difficult. I was only able to think, think, think, "Where did she go?" I could smell her everywhere. This couldn't be real. Everywhere I looked I could see her. Sometimes I swore I could hear her. I remember in the science classes I took we were told many times, energy never dies, it only changes form. So, I went hard with that. That became my first objective. I became obsessed with reading everything I could get my hands on about spirituality. The first two books I read were "Conversations with God," by Neale Donald Walsch and "Imagine Heaven" by John Burke. They both gave me an unexpected view of what death means. I read both within days. My second objective was searching for answers. I started spending hours looking through her room, looking at pictures, reading things I had found. They started to paint a very dark picture. I must warn you to be careful about searching for answers. I don't think I was ready to see some of the things I saw. Being guarded at this point, can be beneficial. Being slower to see and speak really is helpful, as well. I've learned this now. It does a person good to think about what the fall out will be once you have allowed yourself to see or hear something. As I opened my daughter's phone, after the police returned it to me, I learned about many things. Places she had been and hadn't told me about. It was quite obvious she had kept a lot from me as I began to look through her pictures. This alone, threw more salt on my already wounded heart. I couldn't understand why she kept so much from me. I think she feared my reaction or opinions. The depth of the hurt seemed endless. The hurt was growing deeper. As I dug my way through her room my insides felt like they were spilling out onto the floor. I felt so hollow and wounded. So, for those who have to travel this path, take time to think about things. Be sure you are ready to see and accept all you discover. Once you've seen or heard something you never forget it.

I continued to blame myself for all of it. Somehow crushing myself as punishment. I felt extreme guilt for not having seen this coming. Really, how can a mother not have seen this? What did I do to deserve this? Please someone tell me! Why can't I fix this? Please take me instead, let my baby live! My anger became explosive over not being able to change the outcome of this horrific experience. Anything would set me off. Anger with myself for not seeing it coming. I walked around in a fog. My chest felt so heavy.

Days were a blur, feeling sick, and having little will to go on without my daughter. I had experienced so many emotions between sobs of tears. Then, only two weeks later having to return to work and endure some hurtful things that people have said to me. I held back a wrath of vengeance from their ignorance. My advice to those who are reading this would be to turn your ears off. People don't always know what to say to someone who has experienced a tragic loss. You don't owe anyone any explanations.

    I soon joined an online support group for parents who lost their child to suicide, I desperately needed to find a purpose for living again. The group was good for me, in the sense I knew there were a lot of other people going through similar experiences, and because it was online, I didn't need to leave my house which I simply didn't have the energy to do. I had also attended support groups in person but I found it very difficult to commit to and found the online support better suited for me. During the funeral, my friend Patti, had slipped a card into my pocket. It was for a local medium named Deborah Hanlon. I made an appointment. "What could it hurt?" I thought? When I went there, I wasn't expecting much. I even brought my mom, who was not particularly spiritual. My mom told me as we approached her office, "If she breaks out a crystal ball, I'm leaving!" It was the first time I had truly laughed since my daughter passed. Deborah began with speaking about my Cousin Mary's son Danny whom she had lost a few years prior in a horrible accident. The things she spoke of, she couldn't have possibly known. Then she came through with things about Vicki that only I knew. Deborah blew me away!

    In a few months I had heard of another medium not far from my home whose name was Tina Vesely. When I saw Tina, she validated many things that Deborah had also told me about Vicki, things that I had not even mentioned. Tina is a tender soul who was helpful in my healing process. She was full of suggestions, and provided me a different perspective. Tina really seemed to specialize in grief counseling, and provided a spiritual bridge to connect me with my daughter. When I left each session with Tina I was beginning to feel my anger and frustrations starting to melt. I blamed myself and my anger came from my inability to change the past. One of Tina's suggestions for healing was to write a letter to my ex-husband, including all the things I was angry with him about and burn it. Another was to maybe plant a tree or garden in Vicki's name. There were many homework assignments from her to follow over the next two years. Tina worked on reminding me that there were still things to accept as positives, to navigate my darkest days. She encouraged me to have routines to keep Vicki's memory alive and that it was okay to smile, live, and laugh even while I was still grieving. Before seeing Tina, it was hard to experience any joy. I had felt I had no right to smile or laugh when my daughter no longer could. Tina would ask, "Do you really think that is what Vicki would want for you?" At that point, I really wasn't sure about anything

Chapter 5
Vicki's Legacy

Vicki's room was large. As I cleaned her space, I noticed many of her blankets scattered about. I kept two. The other I mailed to her father and the rest I put in a pile. I thought hard about what to do with them. One night I had left work early and noticed the animal shelter across from my job. That night the dogs were barking so loud, even louder than normal. It came to me quickly, that I could donate the rest of Vicki's blankets to the animal shelter. I knew my daughter would have appreciated this. So, the next day I took those blankets to be laundered and gave them to the shelter. I gave some other possessions to her friends. Vicki also had a bearded dragon who I rehomed to one of those friends. Vicki had so much stuff! Over half of it is still in her room now. Another way I celebrated her life was to throw her a small party on her birthday at a local park she loved and frequented a lot during her life. I invited some of her friends to have a piece of birthday cake and we launched balloons in her memory. I bought a balloon for each year of her life and wrote messages on them. Kristi, a young coworker who reminded me so much of my daughter had made some small origami cranes, which I gently tied to each balloon. When we all released them, it seemed magical. Later on, during my second visit with Deborah, she confirmed the balloon launch with no prior information to go on. Amazing, really! I truly believe there are people who can pick up the residual of a person's soul lingering here on earth. I can honestly say, I truly believe in spirituality now. Many things that have happened since Vicki's death are just too much beyond coincidence to be anything but real for me. I believe that there is much more to life here than just what meets the eye! People can live rather blindly if they choose to, but if one is open enough to change their perspective, to try to hear, see, and feel the unseen things, it can open a whole new world.

In order to believe the innate really exists, one must be open to the possibility of more. The mediums I saw gave me hope that all was not lost, and I could still continue to have a relationship with my daughter but just in a different way. For instance, recently I spoke to her through my subconscious. I told her I was struggling and to please remind me of something that would make me smile. Now I hadn't thought about this at all. Matter of fact, I had completely forgotten about this. A vision came to me immediately in the blink of an eye. Then, as if something had brushed against my eyelash, I blinked too. Then I remembered when Vicki was small and we did what we called "butterfly kisses." We put our faces together close enough to have our eyelashes touch. I knew exactly what she was sending me and it made me smile, and in that moment, I could hear her giggle as the small child she once was. I thanked her, as that surely filled my heart with joy. I truly believe I will be seeing her again when I pass. I am sure there is life after death. So, I did the balloon launch for the first two years on Vicki's birthday. But after careful thought, I chose to switch to a more sustainable way to celebrate her life. Every year now, I make a raft with sticks from the yard and tie them together with some twine. I place many beautiful flowers atop, and each year I pick a different body of water to float them away on. In honor of her life, I watch them till they disappear from sight. I think Vicki would have really appreciated this. I look for signs from her every day. I find feathers all the time. Once I found a feather on the driver's seat of my car and the doors were locked! I have an entire album filled with all the feathers she's sent me. Sometimes I hear her voice, other times I find loose change in my path. I have one of her jars filled with pennies from heaven. I'll be starting a second jar soon. I call these little signs "winks."

I leave her birthday, Halloween, Valentine's, and Christmas cards with written notes inside her room. This space remains mostly unchanged. I decluttered it, but I left her bed in the same place, surrounded by many of her beloved things that are hard to part with. I know it's just stuff, but if I were to throw it out it would feel as if I were throwing a piece of her away. I just can't part with these things right now, I'm not ready, so, I don't. Her jewelry, coats, and handwritten things remain where she left them. Her tinsel Christmas tree, I placed in the corner of her room, and this is where I place her cards for her. It makes me feel a little better in this unimaginable part of my life. It's been over five years since her passing. It feels like forever since I last saw her. Moving through all the holidays and life in general without her has been more than difficult. It makes me sad and I still cry often, but I've also learned its okay to cry. Keeping busy and celebrating a loved one's life is a good way to get through it all without falling to pieces.

Chapter 6

Leaning on Spirituality to Survive

I wasn't really especially spiritual before Vicki's death. I dabbled in it with sporadic interest. Vicki and I had discussed the subject and we both agreed there has to be an afterlife. It seems impossible that we come here, and are born for no reason, other than to reproduce and die. There seems to be more to life, learning, and understanding. My belief is that all our life lessons from our previous lives will be brought to our next life so that we will achieve a higher level of self. Understanding the universe and the way it works, the belief and awareness of a soul's purpose, and the journey of the soul. All that are living continue the energy of this world. An immense amount of energy, for sure.

Many of the strangest things have happened since Vicki's passing. I keep paper and pen by my bedside. I do this to write down every detail of when she visits me in my dreams. When I wake, I write them down immediately. These dreams are extreme and different from usual dreams. I know she's been near. Sometimes I can even smell her hair or her perfume as if she's just left the room. I had started a journal before Vicki's passing but now, I'm more attentive about details. Things in the house have gone missing as well, only to turn up weeks later in a place I would never have expected. She used to take things and not return them when she was alive, she seems to have fun with it now in the afterlife. It makes me crazy but she makes her presence known to me this way. Other times, the ceiling creaks, as if someone is walking on the floor above me exactly where her bed is located. There have been times I have heard the back door close. Her bedroom door is near there. Others have also witnessed these occurrences, so, I know it's not just my imagination. There was an occasion once when I was driving alone in the car. Suddenly, I heard this loud ticking. I thought something was wrong with my car. I began to look for indicator lights on the dashboard, with a feeling of dread. But then, music started to play. My phone broke out with the Michael Kors Smartwatch song with Martha Hunt and Zandaya. As I listened to the words I began to cry. They sang about a person burying their secrets in an old tin canteen under an apple tree and to take a swig or three if you want to know their mysteries. The song also describes how that person will never let go and affirms they know everything the other person did was for them. I had seen a medium after this occasion in Kingston, New York, who told me I related to a song that Vicki had sent to me. He then said he didn't understand this, but he was being shown an image of an old gnarly apple tree. I nearly fell off my chair! I also understood that she had sent that to me. It was representative of how she would be here in life! On another occasion, while driving, came the song by Phil Collins, "Follow-you-Follow-me." It played as I was talking to her out loud about the trip I was to take to England with my father the next day. I was asking her to come with me. After listening to the lyrics, it seemed her reply was clear. She would be coming along! She often sends me gentle reminders that she's still around. It's really odd because at first, I didn't realize or recognize these signs but it did eventually come to me. My Oma passed before my daughter. Right after her funeral I sat on my bed with tears streaming down my face, I turned the television on and an Oil of Olay commercial played and in the background was Johnny Cash singing "You are my Sunshine." I sang that song close to her ear, softly, a week before she passed. I have never seen that commercial since. At that moment though I knew she was sending me a message that she was okay. After that I began to notice a sequence of numbers all the time. Everywhere, I saw 1:11 or 11:11. That sequence showed up constantly. I thought it was my lucky number or something. When Vicki passed, I began to see a series of ones and sevens. Sometimes now, they come together and separately. I know deep within my core they are sending me those numbers at the precise time I will take notice. It assures me that they both are near always watching and guiding me. This happens all too often to be coincidental. I have seen some convincing evidence. Sometimes you just have to trust your intuition.

Chapter 7
Resolution

I hate to say this but life will continue whether we want it to or not. Whether we like it or not. Really, the only other thing one could do is leave this life as well. I had thought about that a few times too, but I still had a will to live. I'm now in search of the purpose of life. Why are we here? I've always had a deep desire to help people. It makes my soul feel good. Even so, I still have my dark moments. It's very hard to get through the holidays. I make plans and stay busy, even though there are times I'd rather settle down in my pajamas and have a good cry. I do that anyway, sometimes. It's hard to really let go and smile or laugh. How can I, when she can't anymore? So, I live for my daughter now because I know she is near and watching. Possibly, at times, living through me to visit things that maybe she misses occasionally. It can't make her feel good to see me sad and depressed all the time. I talk to her every day. Some people don't understand or maybe think I'm crazy for that, but it doesn't bother me.

I talk about her often. I feel when I do, it keeps her closer to me. I don't care if people are uncomfortable with it. It makes me feel good and I deserve that much. I've seen the bottom of the bottom of how badly one can feel. It's not pretty. I celebrate each of her birthdays, rain or shine. I still buy her cards and small gifts. I still ask her to stay close and visit often, especially in my dreams. That is where I can actually see her. The only real help is doing what makes you feel better even if it brings a tear or two. It helps to slowly begin to find your new unwanted normal. It's not much of a

How Did I Miss That????

life, to live in a constant state of sadness, depression, or anger. It's been baby steps, but I try to find some happiness in my life now. Not to let her death define me. It can be comfortable to stay in your misery, but it has a tendency to own you. The bottom line is you will now begin a new chapter in your life and it will be hard to navigate. Surround yourself with as much positive energy as possible and do what makes you feel good. Argue with no one about this subject, case closed!

Chapter 8

Helpful Suggestions in an Unimaginable Situation

These are just my thoughts, the ones that have helped me. The best thing I can say to someone who has had to live through this horrific and life changing experience is to not push yourself into anything you're not ready to do. You don't owe anyone anything. Baby steps are in order for a while till you can go further. Make sure you're taking care of your basic needs before anything. Sleep, hydration, nutrition, and taking time for yourself are essential to healing. Don't internalize your pain and cry if you need to, talk if you can. If you hold it in, it will only eat at you. There may be times where people say the most inconsiderate, ignorant, and blatantly rude things and you must ignore these comments. These people more than likely have never experienced such a loss and may not be able to understand or empathize with you. In the beginning you will be numb, bewildered and raw, not knowing how to move forward. But you will have things to do, necessary things that are vital to your time here on earth. You may have other children, a spouse, or other family members and friends who need you. Perhaps these things may not matter to you right now, but above all else, you must carry on your child's legacy. You may need to send someone to the store for you, or have someone help you with small tasks. Simple things can feel overwhelming. If possible, don't force too much on yourself. You may not have the energy to bring yourself out to a grocery store and function with people zipping around you. It may feel overwhelming. A support group may be helpful, even if it's online. I didn't think sitting with a bunch of people, crying, would help me much, but in the very beginning it was a useful resource. I found a website named "Losing a Child to Suicide." There is also another named "Compassionate Friends," I visited these sites for the first six months. I personally liked them because I didn't have to leave the house. Leaving the house seemed to be the worst part of my day. I didn't want to feel sunshine on my face, I didn't want to engage in any conversation with the public. Hell, I didn't even want to be breathing, but I was. Having to see others was something I really did not want to do. I didn't want to get dressed, clean, cook, do laundry, or even feed the cat. This is why baby steps are important.

After only two weeks, I had to return to work my overnight shifts. I cried most of the night. I couldn't sleep much. I would wake after only a few hours. I thought many times, surely the pain would've killed me by now. After only two weeks the freshness of the shock was still there. I wished I could've taken more time off. Two weeks was really not enough. My co-workers were really such a big part of my support and recovery. I talked about how I felt and they hugged me often. It is good to talk about your child. Try not to blame yourself. I did and it really is not helpful. It only makes you feel worse, if that's even possible, and it accomplishes nothing. Finally, you will start to understand that you did the best you could with what you had to work with. Accept tears as your friends because they are not a sign of weakness, but just the opposite. Tears are an expression of the love you have for your child and the strength you have within to continue to carry on.

Even now I still cry and wake with the pain in my chest. Sometimes you will wonder why. It's not easy to maintain under such heartbreak, but you will continue to breathe. Slowly

How Did I Miss That????

I found my way and eventually I began to feel another life starting to form from the shattered pieces of my former life. I have my good and bad days. Grief has a way of raising its ugly head without warning. It can blindside you. There will come a time when you may start to develop other symptoms from loss such as Post Traumatic Stress Disorder, otherwise known as PTSD. I realized this soon after witnessing my child pass. Many thoughts have crossed my mind about this. If I had not witnessed this, would I truly believe she was really gone? Or would I somehow think she was being held against her will, or maybe just ran way, or on vacation somewhere she really loved and chose not to return. The mind can take many journeys on its own if allowed. There's no quick fix to this, only getting through each day and trying to find those tiny embers that still keep your soul burning, breathing, and living. Small, deep inside, barely there, but they are there. One has to look inside oneself. A necessary soul search begins because everything that made sense before doesn't any longer. I find helping people to be most fulfilling and healing. There are people who won't accept help and I have to respect their boundaries. Most times though, I find thoughtful younger people usually appreciate and are grateful for help given. I think it makes them feel good as well. I've been fortunate to find many young compassionate people. Older adults have typically experienced more challenging times throughout their lives which can leave them a bit jaded. On occasion, I meet someone who is on the same path I am. Somehow we are drawn to each other by our loss, and communication is really not necessary. It's a pain only another who has experienced knowing intimately the crushing feeling we live with each day and trying our best not to let it affect everything we do.

Communication promotes healing. Talk to those who are empathetic and to as many who are willing to listen. Try not to just speak with only one person because it may weigh on them heavily. Talking, though essential for the process of healing, cannot bring back your child, and you must find other ways to cope with this heartbreaking loss. Another thing you can do is to think about what made your child happy and continue it for them. Launch a balloon occasionally with a message attached so that their soul will see it. As you release the balloon, call them by name, tell them you love and miss them. In your home, place pictures of their smiling faces everywhere. I kept Vicki closer this way. Listen to their favorite songs, watch their favorite movies, read their favorite books. You can keep a garment they loved and wear it or even sleep with a stuffed animal they cherished. Do what makes you feel better, whatever that may be, as long as it's not detrimental to your wellbeing.

When I returned to work, I had a very hard time functioning. I looked for ways to provide a sort of lifeline for myself. I could talk to many people, but after a short time they seem to avoid repetitive conversations about uncomfortable subjects. Especially if it involves the death of a loved one. So, I put a beautiful picture on my phone of Vicki and every night I kept her on my table as I worked and I found that helpful. Then, I had known that a woman who worked the early morning shift had given birth to a girl shortly before Vicki died. Her name is Joanne. The baby lived a short time but didn't survive. Joanne got a few precious hours to hold her newborn before she would return to what many call Heaven. I felt a connection to her. We both have been joined together by this deep loss and understand the pain it inflicts on the soul. I talked with her often, as I knew she understood exactly how I felt. I cried a few times to her. She was really kind and supportive. It made me feel good to have someone who could relate to how I was feeling. We spoke often about our children and discussed the validity of mediums. Joanne told me one time she was outside with her other two children and this beautiful blue butterfly followed them around for part of the day, somehow seemingly not wanting to leave their side. Joanne was absolutely positive that her newborn daughter, Rhyanne, was coming through to her. It was comforting and even more, very exciting that the butterfly was there to stay. The most amazing fact however, is that Rhyanne's father's sister has a tattoo identical to the butterfly that visited her that day. I don't believe that this is coincidental. I'm absolutely, to my core, convinced that there is life after death.

Chapter 9
Outlook on an Uncertain Future

I wish I could say things will get better. Well, they do a little, eventually you can lift your head off the pillow and your tear-stained cheeks will gain some color again. But the heart, now that's something else. That will never feel the same. Time doesn't take the edge off. I've found at times I'm guarded in life. A thin skin covers the emptiness you will try to ignore. There'll be times it will bleed again. Sometimes from insensitive people, sometimes when I'm alone. It triggers at certain moments. My pain comes out while driving in my car. I worried at first that it may affect my driving, but I can navigate just fine while tears flow from my eyes. My friend, Frances, said it best. She told me "Cindy, we need to move forward now, not back. We've spent long enough in the past. We have to make the best of the life we've been given, letting go doesn't mean forgetting or pretending. It just means learning to live our newer life without our loved one." Frances has a way of putting things straight. I love that about her.

I find nights to be the hardest. I'll see something on television or maybe just a thought crosses my mind that will trigger an onslaught of uncontrollable sobs and flashbacks. This is part of the grieving process. Normal, so they say. Every person I know grieves differently but losing a child is in a category that stands alone. Definitely not the normal process one expects from life. One always assumes that a parent should go before a child. The support groups taught me that

How Did I Miss That????

I was not alone and that many parents were experiencing this similar loss. The stories they share were different, with varying circumstances. Some of these parents sought help from therapy while others didn't. Some took medication and others didn't. A wide range of details, but in the end, it was the same outcome. We lost our child and were grieving. All of us, looking for answers to a multitude of questions that would never be answered. I felt robbed of what I fully had the right to expect from life! Watching my child grow, leaving her all I had worked for. I wanted her to have some things I never had. I thought about her getting married, maybe experiencing grandchildren, but life can be cruel and deny you in a moment. I have accepted these facts are lessons I chose in my life's contract before I even came to this world to learn, in order to make my next life better. This belief may differ amongst varying religions. Each of us make a life contract with our creator. It is because of these lessons I know I will have a very important and fulfilling job when I move into my next life. Should I choose to be born into another life, that'll be quite a decision to be made.

If I have completed all my experiences and lessons, I will then be a master and likely to become a teacher. My hardest last lesson, I believe is to live happily. Finding happiness even in the face of what life brings us here on Earth. I believe this is the hardest thing to do above anything else. I don't think I will choose to come back to this Earth. I believe I've learned most of the lessons I came here to experience, except happiness. I think I saved the hardest one for last. Of course, happiness is an important one, so, I must find it before I pass, or I will have to return to learn again. I am living currently, so that Vicki can live through me as I carry out her legacy. I know I will see her again and she will be proud of me. I look forward to seeing her again alongside my Oma, my second mom Fran Dobson, Kathy Bailey, and many other relatives and friends who have crossed over before me. It will be a happy occasion when we are all together again after I pass, myself.

Chapter 10
Thoughts Start to Change - Need a Retreat

I decided it was time for a change of scenery. My dad offered to take me to England. I decided to go with him to see my sister who lives there. It was my first visit to the U.K. Before I left, I did some research on mediums. I googled the top ten and reached out to several. I also learned the best ones are located in England and Brazil. There is a college just for mediumship in England named Arthur Findlay. It's a college of spiritualism and psychic sciences. It has many courses and was founded in 1964.

So, the day came, and off we went to Newark Airport. My research on mediums gave me the name of one man I wanted to see. I was drawn to him. His name is Shane Gadd. With the help of my sister, Linda, we made an appointment. I asked her if she would go with me while I was visiting. It was one of the best decisions I have ever made. He has brought so much comfort and healing to my heart. Shane once said to me, "When you are in a dark place you sometimes tend to think you've been buried, but perhaps you've been planted, so Bloom!" He boosts my positive thinking which can be difficult at times. Being positive, I've found hard to do with such a huge loss. I feel I've lost myself as well, at times. Shane is always filled with suggestions and helpful thoughts. He also has a heavier hand about setting you straight, so, his honesty is very valuable and appreciated. Some of his suggestions are easy to do. Many are lighthearted, such as a ritual or a thought process. Others require stripping back the layers to look deeper inside oneself, taking stock and doing some real soul searching. All are helpful in healing the heart. He is a great life coach.

My dear friend, Frances, has been a blessing. She said to me on one occasion, "Cindy, we are here to help each other." She's quick to remind me that I've helped her equally. I don't feel as strong as she is though. She and her husband have such strong faith. Our friendship bloomed from a horrible tragedy. Losing someone to suicide is brutal. When it's a child, it can be crippling and life altering. I always say no matter what type, a loss is a loss. Fran and I know however, the loss of a child is far more complicated. Especially when it comes self-inflicted. The fall-out leads to other complications for the bereaved. Fran always says that time is making us a little stronger, bit by bit. I guess she's right. She's so positive and that positivity is a key to survival. It is important to surround yourself with positive people. I wouldn't advise anyone to try to recover from this type of loss alone. You really will need help. You will need people to remind you, you did the best you could. I was not a mind reader and I am human. Yet, still many including myself, tend to put the entire blame on themselves. I beat myself up for four years. This year I finally accepted maybe I wasn't entirely at fault, however, I still feel waves of guilt.

My sister, Linda, delivered an essence of spirituality to me on Angel wings. I always believed that a person's soul doesn't die with their body. As I said here before, I thought about it occasionally before Vicki's passing, but of course, much more after. Linda slowly reminded me, and brought all those beliefs back to me in her own way. Unbeknownst to me, she had been studying the innate and spirituality for many years prior. This brought us closer, and at a time when I needed it most.

How Did I Miss That????

Funny how things work sometimes. Linda gently guided me towards continuing my relationship with Vicki. It was just in a different way. This alone, was a very comforting thought. I was and still am broken. I was drowning in my own sorrow, rawness, guilt and endless emotional turmoil, just waiting for the pain to kill me. Even the smallest belief that if I shifted my mind to an altered state of consciousness then I could continue to hear and communicate in the slightest essence of my daughter excited me. Changing my own thoughts to a distinct range of vibration gave me so much hope. I just needed to know how to sharpen these skills in order to do this effectively, and so, meditation began for me. That was the beginning of a whole new way of thinking. I was awakening in a way I had never before. I tore down all my prior thoughts and beliefs, I mean, all prior beliefs shattered. All to enable me to heal and become whole again. It would be a journey through hell but it carried the hope that I would reach the other side of a beautiful life once again.

How Did I Miss That????

Vicki's death has changed many people. I'm sure some of them still think of her from time-to-time. Her friends used to come around often, but not as much anymore. People told me this would happen, however Christina, still comes around. I enjoy seeing her and spending time together. She has become such an important part of my life. She remembers things Vicki did and spoke of. We talk about her often and it fills part of the emptiness within me. Christina's and Vicki's friendship began in the sixth grade. They had a lot of history. Christina and I now do many things together in Vicki's honor. I believe Vicki feels this. Christina, being such a sensitive soul, notices the smallest details, like a tiny feather while walking along a pathway, or a passing butterfly. Always being quick to point it out to me so that I don't miss a thing.

Linda told me once of an occasion that came upon her. She described to me an average day in which she was driving to a destination that seemed forgotten, when she came to a stop as the traffic light turned red. That was when it hit her. She seemed to slide into another level of consciousness, a heart-to-heart soul connection. Her body no longer felt bound by constraints, skin, separated from others, seemingly weightless. It was a warm and loving place, fully supportive so that she could be safe to just exist. The noise of the traffic and radio faded away, as peak awareness came with the realization, she was in a state of being in oneness. Her eyes fell to the long grass waving in the gentle breeze and in her mind, she understood the words, I am here, in every blade of grass. Linda then saw something that drew her towards a bird flying overhead, and Vicki popped into her awareness. Vicki always had a love for birds. Linda went on to say, in that moment, she felt Vicki's presence, in the bird, in the grass, in the sun as its rays shown down from the sky. The words, "I am here" popped into her brain again and she knew that my daughter was with her in that moment, just like having another

How Did I Miss That????

passenger with her in the car. All this happened in the time it took for the traffic light to turn from red to green. But those few precious moments will last a lifetime for her. After our short visit to England, once again my father and I were on a plane to return home. As I entered the door to my house somehow, I fully expected to see Vicki there. In reality I knew she would not be, but as I stepped into the open emptiness, I couldn't stop the sadness that creeped inside my head and through my veins. Back to reality, sadly.

Chapter 11

Regrets

After these past five years since Vicki's death, I think a lot about things that I didn't before. I think about a lot of time I now consider wasted. By wasted, I mean, time better spent on things I really wanted to do. Places I really wanted to be and with who I really wanted to be with. Was it really wasted? Hindsight is always twenty-twenty. People think at times, what they are doing currently is what they should be doing, not really considering the importance of what they are not doing because of it. But if they knew what would later happen down the road would they regret having done that thing instead of the other where their heart really was? Would you consider that wasted time? I have many regrets. One of them was how much time I spent working. I worked like a robot. Snow storms, if I had a fever, if Vicki had what I considered unimportant school functions, if my Oma was lonely, I still went to work. I didn't like making executive decisions about buying a pizza for dinner. When my husband and I were first married I used to get mad that he worked overtime. But now that we were a little older, owned a home, and had a child, I expected he would help by grabbing some overtime or maybe a small side job but that never happened. That's why I went to work and because he worked nine to five I had to work both evenings and overnights. My ex told me the bills would get paid when they got paid, period. Vicki was growing up quickly. I hated buying her clothes at yard sales and thrift stores. Kids can be so mean in school. Vicki needed many things and owning a home required constant repairs. Cash was always in demand. The water heater broke, the washing machine gave out, the furnace needed fixing, and we were always short on the funds for the mortgage payments. It was very stressful for me. Writing the weekly bills out nearly gave me a nervous breakdown. I had no choice but to suck up as many hours at work that I could get.

I regret that now. Vicki is gone now and it made no difference. Maybe what would've served her better was quality time with me. If I had, during her childhood, maybe she would not have developed the depression she had. Maybe we would've had better communication. But this is the maybes, would've, could've, should've that I spoke of earlier, that will haunt me for the rest of my life. I regret not spending more time with other loved ones who have since passed. In reflection, my job was so unimportant. Time is the most cherished gift that we, the living, receive. Once spent, you cannot have it back. So, my advice is to use it wisely. Always surround yourself with the people you most want to be around. Your loved ones, because when you say yes to one, you are saying no to the other. Be sure to make your time count and seat those at your table whom you really wish to spend your time with. Don't waste time doing things because you feel obligated. Only your heart will be the truth teller on which choice is most important. Don't waste time doing one thing when you'd really rather be doing something else.

I really don't know where the next phase of my life will take me. I've been forced into this new chapter but I don't wish to let my trauma own me. I don't want it to define me, but sometimes I feel it does. I don't sleep well now. I have chest pains from time-to-time, and feel anxious. The rain brings pain. I experience many symptoms of PTSD since Vicki died. Sometimes it can be extreme. My wish is that maybe, somehow, it will make me a better person. I try to do more of what I enjoy and spend less time with people and doing things that require too much of my energy. Of course, my work isn't something I particularly enjoy anymore, but the reality is I can't expect to live for free, so, I make small changes until I can make bigger ones. For instance, I enjoy helping others. It comes from deep inside and my heart tells me it's right. Kids and especially young adults in particular. I think this is becoming a major part of my new purpose in this life. I give a lot of thought to my future. My plan for the years after retirement is to fill my time with doing things like photography and volunteering. Maybe painting more pictures, which I haven't done in years. Perhaps camping and enjoying nature or writing more books. I really give it a lot of thought and want it to happen and so it will be. It's really about what drives oneself. All the things that have long been put aside. Even new things never considered before.

    I do try to spend more time outside to reconnect with nature when possible. There's a sense of freeing one's soul there. It may not be for everyone, but it works for me. My biggest obstacle is anger, which is part of loss. I'm angry often, for many reasons. For things I did, for things I didn't do, things I accepted and shouldn't have. Things I can't change, things I had no control over, hurtful things I've said to people, and hurtful things others have said and done to me. I regret not sticking up for myself more. I am quite a sensitive person now. I get upset with myself for not spending more time with my daughter. Not recognizing her sadness behind her smile. Some others that I should've spent more time with too, and missed that opportunity, which I can never have back. Some people I cherish are still here, so I put more focus on them, try to make better choices, and make my time count for more. I wasted so much time with men who didn't appreciate me. Waiting for things to change, waiting to receive the love I deserved but never had. So much time was wasted. I needed to learn to love myself. It took me so long to realize that. I wasted so much loyalty on employers that had none for me. The list goes on. I've been told in order to heal and let go of my grief that I would need to let go of my ownership of it. The anger that comes with my grief sometimes fuels me. I have refused to let go and I release that anger in spurts, letting the pieces fall where they may. Not even trying to contain it, simply not caring. At times, feeling its familiar comfort. I know how to be angry. A great statement of myself and how horribly my heart has been wounded. If I let go of this anger, what will happen to me? Anger and my grief go hand-in-hand. If I let go of the grief and anger, does that mean I forget my daughter?

The answer, of course, is no. But it's a different way of thinking that guides you in a new direction. A way out through the other side, to the "new normal." Instead of concentrating on how I wish I had had more time with my daughter, and how her life ended much too soon, I instead try to remember all the detailed moments of the twenty-six years I did have with her. If I could release the anger, the pain might be less. It may even bring healing, maybe I'd even be able to start forgiving myself.

I try to put some things in place to guide me so I can travel that path when I'm ready. Moving forward is not easy. I am not fully ready yet. That horrible day isn't far from my memory. It really doesn't seem that much time has passed except when I think about the last time I saw my daughter's pretty face. I guess happiness is a thought process. A way a person feels about having done something. You experience this feeling every time you do something kind for someone. Sadness and depression happen when you get intertwined with guilt, anger, grief, hopelessness, fear, etc. Those who have lost a child to suicide often will take ownership of these things as I have. You must experience these emotions before you can release them and this is part of the grieving process. However, it is crucial that we focus on the positive emotions and good memories in order for happy energy to surround us and lift us in a desperate time. It is a painfully slow process. You can immerse yourself in it. You can breathe it in, you can. Everyone grieves differently. There is no set time scale. Only the individual will know when to take the next step in the healing process. Of course, it doesn't stop some people from thinking they can put a time limit on it. Don't tolerate this under any circumstances, not even for a second. I'm told happiness will always be there waiting when I'm ready to accept it. I look forward to that day, but I'm not quite there yet.

Of course, I have good days and bad days. On good days I still feel like a million pieces shattered. Even after feeling so broken and hallow, I still go to work, still help my parents, still hold conversations with people and am still somewhat responsible. However, most times I feel I'm on auto-pilot. I have been working to change that recently. Learning to forgive myself plays a major part in this. Once I totally forgive myself, my soul can truly be set free and experience the happiness of life while I'm still here. I feel then I will finally have conquered my final lesson.

Chapter 12

Forgiveness. Is It Really Possible?

I'm still trying to find ways to forgive myself. How is it possible to do that? But somehow, I still try. With the help of family, friends, Tina and Shane. Slowly I am beginning to find purpose. Tina Vesely first, then Shane, had given me tasks I mentioned earlier. Tina gave me homework first. She had me write on sticky notes all the things I needed Vicki to forgive me for. These messages filled a wall in her bedroom. She also had me write letters to my ex, expressing all the rage I had for him and why. I wrote down all of it and burnt them in a fire in my backyard. Tina explained to me her understanding of what happens to souls after the body dies. I had so many questions. She provided so much knowledge and comfort. Currently, I continue to Skype with Shane. He also played a large role in where I am today. I still have a long way to go, but he has been a beacon of light through the darkness in my healing process. Helping me find my purpose and finding the answers to the vast number of questions I still have. Shane is a holistic medium by nature and has also pointed out the different ways in which the innate communicate with those of us still here. He explained his understanding of how souls are connected and how one can recognize some of their signs. I don't wish to change anyone's beliefs here; I just want to share my own story so others can maybe find help in it. Perhaps something can be learned from the most desperate, painfully raw time in my life.

How Did I Miss That????

I'm still trying to learn forgiveness. Mostly for myself, but for others, as well. Every time you think about something awful someone said or did to you, you allow them to hurt you one more time. Forgiveness is really the only way to release all the negative emotions, but it is not easily attained. Vicki's father told me the night I called him from the hospital with the news of what happened, that Vicki took her life because I had treated her like my mother had treated me. It was one of his last attempts at hurting me but perhaps it was his way of expressing the pain he was feeling from the loss of our child. There came a time I had to forgive him as a means to move forward. Vicki and I had a tumultuous relationship towards the end of her life. We had arguments and disagreements about daily life but that didn't mean I loved her any less. I thought as she grew older that would inevitably change as it usually does with children and their parents. As a young adult, I didn't always get along with my parents but that is the case for many. Of course, it is something I grew out of as I aged and I hoped that as Vicki matured, we would become closer and forge a stronger bond. There was not enough time for that, unfortunately. On the flip side, I am grateful that I had the time before my mother's passing to make amends with her. My best advice is to forgive only when you're ready and don't be afraid to accept empathy from those who can genuinely give it.

Chapter 13

Looking Back over Vicki's Twenty-Six Years

Sometimes I think, I kind of knew in my heart of hearts that Vicki would not be here long. My intuition knew it even when she was small but it's something the mind denies and covers. Certainly not something a mother would think about. What a horrible thought. I wish though I had considered that it could happen because if I had listened to that tiny inner voice maybe I could have tried a few things differently from the route which led us to this day. I have many regrets about this. More of that second guessing of myself.

My only salvation now is to forgive myself. There aren't enough candles in the world for me to light, so it must come from an inner strength. I know I made many mistakes. I must accept I'm human and I do know all the decisions I made while she was here were for her safety and well-being, even if they weren't exactly right. My biggest fear was that she would get hurt. Vicki lived so fearlessly. I never dreamed she would really harm herself. I knew she struggled with sadness but I believed that she'd be okay.

After she left, I found something she'd written prior, which opened my eyes to the depths of her mental state. She really hid it so well from me. I was never one to go snooping through her room. At twenty-six I didn't feel that was right. If I went into her room, it was only to empty the garbage pail. I still blame myself for the unseen. How do you blame yourself for not being told or not seeing a problem? You can and do though. Life moves at lightning speed. Many people, especially single parents in particular don't always have time to stop to smell the roses from time-to-time. There will be many people who will say, if it were my child, I would've noticed. Not necessarily I say. But many people will say the most unbelievable, untrue things to you in order to validate that guilt. This is the lack of empathy I don't understand. I hope I was not one of those people before Vicki died. I seem to be more accurately aware of how words can come out so hurtful now. It tends to be so hard for me to remember my life before she came into it. I have however decided not to give these bold people who feel they can say whatever they feel to me without consequence, an ounce of space in my head. I would surely say to anyone suffering this same trauma to do the same.

Chapter 14
Triggers and Tailspins

Triggers can come out of nowhere, likely to happen, and very unexpected most times. Sometimes not though, making these types easy to avoid, which is good. There were times where an event would trigger an unexpected flood of emotion. For example, a friend of mine, Whitney, was having a baby and asked me to come to the hospital for support. She said her mother would also be there. After her call, I drove to the hospital. It was the same hospital where Vicki died. I sat frozen in my car. I had to call her and tell her I wouldn't be able to come. Whitney knew about my daughter and understood completely but it was such an unexpected reaction for me. As I sat there looking around a flood of images started to form. The entrance where I went in that night, where I came out, where I dropped to the ground sobbing and howling to the moon in pain. It all started to come back, I just couldn't step foot in there to welcome this new life, when my baby had died there. I just couldn't. I've already established I suffer from PTSD. Until you experience PTSD, most people don't understand how debilitating it can be. There are things that will happen in life that many people can roll off their backs and it just doesn't seem to faze them. But when traumatic events occur this becomes impossible, stopping some people in their tracks for years. In a death from suicide, many occasions can become intertwined with that loss. Birthdays, certain places, holidays, and the date of their passing can be very emotional. A feeling of tightness in one's chest and edginess can start even days before these occurrences. I didn't even realize why I felt this way at first.

How Did I Miss That????

On one May 31st, not feeling well, I lit a candle for Vicki, said a prayer for her and went to bed. I woke at the exact time my daughter died. As I entered the bathroom in the darkness of the night, red and blue lights were flashing as they lit up my entire bathroom and the street on that side of my home. An ambulance was parked in the road. I felt the panic and fear setting in immediately. As I looked out the window, I saw my neighbor, Frances, being brought out of her home. I rushed back to my room in panic mode. "What should I do?" I repeated to myself. I threw on some clothes and ran out the door toward Fran's house. I was blinded by the flashing flood of lights. It was very difficult to set aside the fear and flashbacks of that night in order to focus on making sure my neighbor was okay. As I spoke to her, she told me that she had called the ambulance because she felt very unwell and dizzy. Fran was so frightened. I told her not to worry and that I'd lock up her home and call her friend Anita. I spoke with the ambulance crew and went home. I returned back to my house and sat, safe but rattled. I grabbed my head with my hands and curled into my knees in tears. I said another prayer for my neighbor and asked Vicki to look after her. Thankfully Vicki heard me, because the next morning Frances was home and feeling a bit better. Frances told me she didn't remember much about our conversation the night before. I asked her what the doctors told her and continued to check on her after to make certain she was okay. It literally took me two days before I started to feel better. Seeing those bright flashing lights again shook me to my core. Bringing some of the emotion with it. Feeling scared and helpless. It all seemed to freshen the rawness and slightly paralyzed me again. My initial thoughts were "No! This can't be happening again!" That fear grips your entire body, making it hard to breathe but there is no escape. The helplessness is overwhelming. The only small bit of control I had was to be able to focus on Frances. Now I can see the difference between my old careless normal to my new normal. Protected and guarded. I will never be the same. How could I? The only good thing I can say is that time has softened the rawness of the wound that is inflicted on my heart.

Those who have lost children are able to eventually function once again. This is our new and unpredictable normal. The triggers that send us into tailspins often come without warning. They come on fast and cripple you in a moment. But we will wake and live another day because that is what our child would want. They certainly wouldn't want us to linger in the sadness for the rest of our time here, but rather maybe live life through us in a more beautiful happier time. I believe that in the afterlife there is only love and it radiates to us here on Earth.

Chapter 15

Afterthoughts

There were times while writing this book I had to put it down. Remembering the details, brings the memories back to the forefront making it hard to carry on with everyday life. The memories of the pouring rain, the lights, the police and the paramedics, the unexpected television reporters who rang the doorbell the following day, the realization that something permanent and horribly wrong was happening in front of me. The hospital, the waiting, the trauma that followed, the funeral, the shock and confusion, trying to make sense of something that made no sense, wanting to change the unchangeable, accepting the unacceptable. All brought with an ocean of tears, frustration and anger on an emotional rollercoaster. I demanded to know what I did to deserve this. Why was this happening? What had I done so awful? Could I take it back? It has changed the way I think about almost every detail of my life now through both good and bad days. Everyone says they look for closure when they lose someone they love. I have not found the term closure to be accurate. I don't believe anyone who loses a child to suicide has closure. This is my own opinion, but I believe it's a fallacy. It is next to impossible to move on with one's life once this happens. How can you? It's like leaving your child behind to move forward. But I've come to a conclusion, and that is I am not leaving her behind, rather, I am bringing her forward with me. Her DNA is part of me. Therefore, I believe her soul can see through my eyes. In this thought I try to show her a better life than the one she tried to escape. I pray for her to experience this happiness through me. I'm not always successful, but it is my newly gained spirituality breaking in to save my soul. It's a slow journey. A slowly healing journey.

My daughter does travel this journey with me and remains nearby always. She is subtle and patient, always guiding me. Knowing me like no one else. She sends me little winks, letting me know she's there. I often hear her bedroom door close. My daughter has sent me hundreds of feathers. She's left them inside my locked car, inside my work place, on my porch, in my kitchen and a lot of other places one would not expect. I find loose change everywhere. I know these pennies are from her. Sometimes in the stillness of a moment I can feel her presence. I feel her touch my hair, my cheek or my arm, or the hint of her scent randomly permeating the air. It is hard for people who have never experienced this to fully understand. There are times that come along this path when you will meet someone who has suffered this same tragic loss. Talking with these people, the words can be few, but the understanding of the heartache is a very precise one. It is true, that eyes are the mirror of one's soul. The pain is shared intimately between these people. We are drawn to each other. Often without any confirmation. I know our children put us in each other's path in order for us to lean on each other. Sometimes in a weak moment supporting what seems like the entire weight of the others being. But we do it for each other because the intensity of this pain never stops. Losing a child is life altering. It's like coming home from a war, battered, bruised, and suffering from the inside out.

People who suffer from PTSD have invisible scars. We look normal but are far from it. Minute things can trigger a landslide of emotion. So, please, use care when speaking. You never know a person's story. Kindness is in order for those like me and others who suffer this type of tragedy. To those who are in my shoes, I wish all the peace, love, light, and healing to you that your soul will accept.

Chapter 16
Wishes

There are many things I would've done differently given the opportunity to turn back the hands of time and still retain the wisdom I have developed. Accepting things that happened the way they did and my life as it unfolds now, is and has been difficult. However, if I was given three wishes they would be that Vicki has found the happiness she was looking for, for myself not to have gone through the trauma my life has brought upon me, and for world-wide happiness. It would be amazing to find and live in happiness. The world is filled with so much judgement and hate. I know I must find some happiness in my life before I die or I will have to return here in my next life to learn that very important lesson. I feel I have lived many lives and saved the hardest lessons for last. The final lesson being happiness throughout life's storms. Living happily even in the face of death, disease, war, betrayal, the infliction of physical and emotional wounds, and poverty.

I don't feel I want to return here again, rather resting on a different plane of the universe and experiencing that as a master. I feel I would be more useful teaching or helping others who struggle with the different lessons they could not achieve, as they arrive at the vibration we know here as heaven. A person once asked me if I could bring Vicki back with one wish, would I? In the beginning I would have said yes. Without a thought or hesitation. Of course, I know that could never happen. But now I ask myself would it be selfish to bring her back for my own wish when I know she desperately wanted to leave? I believe I would choose to bring her back only if I could be promised she would be happy. As for now, she will be forever twenty-six to me.

How Did I Miss That????

    The only one thing I know for sure and probably always will, is I miss being able to touch you, Vicki, talk to you, smell your hair, and see you with my own eyes. It's the physical part that is the hardest to get past. Knowing you are always here with me is a great comfort and I have allowed myself to truly know this but the physical touch of a human cannot ever be replaced, and that is what leaves me broken beyond words. The final words I have is for my daughter. "I miss you and I love you now and always will. Please stay near to me and I will see you, soon, when my time is due, darling."

Chapter 17
Visits from Vicki Through the Dream State

My Vicki probably gets very tired of hearing me ask for another dream visit. I can picture her saying "You know...I'm quite busy here now, mom" with her hand on her hip. It makes me laugh because of how sassy she could be but I still ask often anyway. She will eventually come through. I can tell the difference now between my ordinary dreams and these special visits. They have a different feel to them. I cherish them and write each one down. Every time I experience one of these dreams, they leave me aching for another but I know I need to learn to live without her. She's gentler with me now and when I really need a visit, she seems to know and comes at the perfect time. My first dream came on the eighteenth day after she left. I wrote it all down so as not to forget the smallest detail. As I reached for my pen, I began to write these memories.

June 18, 2015

I just woke from a dream! Vicki and I were driving through what looked like snow-covered sand dunes. We found our way to a paved road by going down a muddy wet hill. All the while she was laughing wildly. We hit the paved road hard, and the wind was blowing back her hair. As I drove, I turned to her and said "You know you're not here with us anymore, right?" She said "Yeah." I then said to her "I have something to ask you." She said "what?" I then asked "You know how much I loved you, right? You know I loved you?" She said "Yes, I know, ma." I began to cry and pulled off the road next to what resembled a bus stop. We got out and walked to the front of the car. She hugged me and said "These are the worst tears you'll cry, ma." I hugged her back, looked up, and poof she was gone. The dream continued as I then started to think about an accident I had with her in the car with me, when she was still a toddler. On our way home it was beginning to snow and the car began to pinwheel and I lost control. In the spin I woke. I had tears in my eyes. I had been asking for her to come to me in dream form. Finally, she came. It was so real. She was dressed in her normal black and it wasn't cold in my dream either, but the sand dunes were covered in snow. She was so alive in the dream. As she was laughing, she lifted her arms up as if she were on a rollercoaster. When I woke, I could've sworn she just left the room.

February 5, 2016

I was standing in a bathroom. It was not my bathroom. It resembled the shape of an elbow. As I stood facing the mirror in my robe, she came into the bathroom. She was little. Maybe seven or eight. She was in her nightgown and robe too. She was upset and crying. I asked her, "What's the matter?" As I kneeled down to her level, looking into her face, she said she was scared because she was playing PlayStation and everyone was dying. I grabbed her and pulled her into me. I hugged her tight and told her it was okay. "You're okay," and then she was gone.

February 15, 2016

Ten days later she came again. This time she brought Franny, my second mom, who passed exactly six months after Vicki. Let me take a moment to explain all the "Fran's" in my life. There are three, my second mom Fran and neighbor at the house in New Windsor. My neighbor Fran, across the street from where I live now, and Fran, my dear friend who lives in Ireland whom I became friends with through the support group. Getting back to my third dream…Vicki and Fran were chasing each other like squirrels around the tree in her old yard. I say like squirrels but it was more like watching wind. Very hard to describe. I watched their energy laughing and giggling. Round and round her yard finally ending by chasing each other up the trunk of her huge tree, circling it round and round to the top and off the branch tips. Exploding off the tips like sparkles and then floating off like lightening bugs. Then I woke. It was so surreal. There would be many dreams to follow.

Another dream came with a vision of a row of shimmering evergreens, but they were white and sparkling so brightly. All of a sudden, I saw Vicki dart out around the pines and disappear again. Then I saw a boy chasing her. They were running and laughing loudly, chasing each other. Then in a flash someone came running toward me fast and stopped short causing the snow to go up like a cloud of smoke. It was Oisin, Fran's son, who was standing there in front of me! He said "ello right." in his Irish accent. I asked him, "You want to tell your mom something?" I said, "She's having it rough you know." He replied, "Tell her eh, I'm okay, right...and what she been doing in me room? Please tell her to smile more for me." His face had the biggest smile and snow stuck to the tops of his eyebrows. Then out of nowhere someone came running up behind him and gave him a big push and he went off in a puff of white smoke. The smoke was cold like snow. Vicki then appeared laughing playfully loud. She kissed my cheek and was gone. What a dream. Our children seem to be connected because Fran and I are connected. Wow.

October 28, 2016

I had another visitation from Vicki. I dreamt she had called me. I was riding in the back of a car looking out the window. I never saw who was driving. I told her I had broken it off with an old boyfriend, and she said she knew, that she had seen it on Facebook. She said he had written he was giving all his money to charity. Then I told her I was off from work on Halloween and that I was also taking that entire week off, if she would like to do something. She said she thought I'd probably wind up going to work. I told her, "No way." She then said it would be nice. I watched out the window as I rode in the car. I saw snow piled on the side of the road and I woke. I felt so peaceful as my eyes opened.

Halloween was my daughter's favorite holiday so I knew she would be around me.

I also have visitation dreams from other loved ones. One night Fran Dobson came to me. I miss her terribly. She was my rock and cheerleader. I confided in her so many times. I begged her often to visit me. Then finally she did! I dreamt I was laying on my stomach sewing on the living room floor. I broke the needle I was using and when I turned to get up, she was sitting in my recliner facing me. Someone named Bob was sitting with her. He was in the chair to the right. Now this was odd, her husband who had passed years before, was named Bill. Fran was wearing a chocolate brown velvet sweatshirt. I told her I missed her so much and she told me she missed me too. The dream was very short and I don't remember all the details, but I'm good with just knowing she was with me for a night.

My friend, Kathy, had also visited me once shortly after her passing too. Kathy died on my birthday. It was so exciting to have her finally come to see me. As my dream unfolded, we were much younger, the age of school children. We were playing a game of mailman. She had her little mail cap on and was driving one of those plastic cars that are powered by foot. She pulled up to me as I stood on the side of the road. With a big smile she handed me some letters. As I took them from her, she was so happy. It filled me with joy. As I woke, I began to process the dream. Kathy's father was a mailman all his life until retirement! I knew now, she was fine and with her dad. These dreams I have are small gems to me. It's really the only time I can get that close. I can travel uninhibited. Free from the confinement of my body. My dreams take me to wonderful places. Sometimes as I wake, I try to force myself back to sleep to continue the dream, but it never does seem to work that way. I guess like the ending of a great movie, the end is the grand goodbye. But in a dream, it's more like see you again in the next dream visit. Vicki has come to visit me many more times than I could write about here, but each one is like precious drops of gold.

I was sitting here thinking how each year that passes, I seem to grow a little stronger, and have more processing capability. I think the second year was the very hardest to cope with, in the death of my daughter. That probably was because the numbness was beginning to wear off. Another dream I had followed shortly after as I was watching Channel thirteen one night about the way older movies were created such as "Viva Las Vegas," "The Wizard of Oz," "Singing in the Rain," and "Yankee Doodle Dandy." I was beginning to fall asleep and heard the song "Somewhere over the Rainbow" sung by Judy Garland. My dream began as I was sitting on a couch with my Oma and we were watching the television. There were little trees twirling around on the screen and special effects of the way life was being portrayed. It appeared in Technicolor. Everything was in bright oranges, greens and blues. It was really brilliant. Oma kept looking back at me and I said I wish life could always be like that. Vicki was sitting on the floor next to where I was on the sofa. I rubbed my Oma's foot and leg with my foot as we were seated very close to one another. I was careful as I knew her legs were tender. I started to feel tears come because it made me feel so happy. I also somehow knew Fran was there, I felt her presence. This has been the only time I had three people visit me at once. These dreams bring me so much joy. Dreams allow you to look through that thin veil into the innate's world. I continue to look forward to more adventures!

Chapter 18

How to Support someone who has Lost someone to Suicide

In my experience, I have found people grieve in many different ways. Only by knowing the person, can you judge exactly what is best to say. People who have lost loved ones to suicide are really separated from those who have lost those to injury, sickness, or age. Guilt and shame often play a large role. I wasn't one to hold shame, as I knew Vicki's mental state, but some do. I made the mistake in believing that because my daughter was so strong willed that she'd be okay. Suicide often comes without warning. People who choose to end their lives, I have found, often hide their intentions. You could see them smiling today, as you sit and talk about plans for tomorrow, next week, next year, and when tomorrow arrives, you find they have left this earth. Shocked, as you think, I never saw that coming. Worse yet are those who find their loved one deceased which can be devastating and psychologically altering. As time passes through the years, a person is reminded with every birthday, holiday, event, and memory that follows. Many times, others may not know what to say to the person who has lost someone to suicide and that's okay. A hug or just a simple "I'm here for you" is comforting enough, rather than saying something unintentionally insensitive. I realize people often feel like they are walking on eggshells. However, it is important to those who've lost someone to suicide to be reminded that they have people they can count on to be able to pick them up when they feel like they can't go on anymore. I have also found for myself the textbook stages of grieving to be untrue. I found my grief would raise its ugly head on many unpredictable occasions. When the suicide first happens, you are just numb and in shock. You may find yourself sobbing uncontrollably, confused, numb, or forgetful. Sleep seems to be impossible and it is even harder to even just get out of bed. You may also have flashbacks to that dreadful day. The night my daughter died played over and over in my head. By the time of the funeral most of the details were a blur, but the image of that night never changes. They are as vivid today as they were that horrible night. Parents are left to first realize what just happened is not just a nightmare, but in fact real. They have to accept the unacceptable, planning a funeral for a child. This is not the order that this is supposed to happen! Arrangements will have to be made and a good support system is vital. It can be paralyzing having to deal with the aftermath of such a debilitating loss. You may find yourself thinking of the last conversation you had with them and analyzing every detail. Why didn't I talk to my daughter more, hug her longer, or ask more questions? Why did she hide this from me? Why didn't I recognize her sadness?

Those grieving often avoid social situations because they just don't have the energy to talk about their loss. It can be triggering and detrimental to those who are not ready to talk about it. Parents may even consider suicide, themselves. I couldn't even imagine how I was going to continue on when the one person I had worked all my life to protect was now gone. I felt that I had failed at my most important job, being a good mom. I thought that I needed to be punished. However, inflicting punishment on oneself is not going to bring back your child nor is it beneficial in the healing process.

You may start to realize, I'm still breathing. What do I do now? What can I do? Some even start to hate the world. Even lashing out at everyone around them. This is a strong sign to how bad it hurts. This can take years to diminish. It often takes relationships to the edge. Anger sometimes is often displayed as the deepness of the wound never goes away. After a while though, most people tend to see that they are still alive and breathing. And with that, hence the bargaining part of the guilt stage begins. I did experience this. I asked for her forgiveness constantly. I had conversations constantly about what I could do to prove to her I'd be a better person. I asked her to send me signs she wasn't angry with me for being impatient and cross with her sometimes as she grew. I apologized to her out-loud a million times. I promised I'd not make those mistakes again if given the chance. If I could just have one more moment, one more hug, one more conversation with her. In days, weeks, months, you begin to realize it's not going to happen like that. I began to miss the physical part so desperately. I hurt to my bones, my insides feeling like they'd been through a blender. That is when, what I learned from the mediums, was very enlightening. This doesn't work for all but I'd say to everyone, try it once. It freed me from my self-punishment and I began to think about the quality of my life I had left and how I could use it to make a difference. Of course, carrying on Vicki's legacy would play a major role in that.

As I mentioned previously, survivors of suicide loss could greatly benefit from support groups. Many need help with making sense of the death and a better understanding of any psychological problems the deceased may have had. Most need treatment for PTSD they now carry. Often, family members will blurt out things that can be extremely insensitive. These people have never had this type of experience and lack the capability of being empathetic. They simply just can't fathom how deep the pain is. Support groups are good at exploring unfinished issues in their relationship with the deceased aiding in coping with reactions of family members.

So, for those who don't know what to say or do for someone who has just lost a child to suicide and who are grieving. I'll list a few things that are usually safe to say and do:

How Did I Miss That????

- I don't know what you are going through but I care and I am here for you.

- Don't wait to be asked for help. They may be too grief stricken to know what they need. If you are able, offer to grocery shop, babysit, cook, or help with household chores.

- Don't be afraid to speak the name of the person that died. In my experience it is important to the survivors that their loved ones are always remembered. We appreciate that you let us know that they touched your life and you still think of them.

- Sit with the person who is grieving, watch TV together, and be present in each other's company.

- Listen more, talk less, and be patient.

- Remind them how important it is to stay hydrated and nourished.

- Keep reminding them to be gentle with themselves and to rest as much as possible.

- Continue to check-in, those experiencing loss may not be able to reach-out to others. Let them know you are available to them for help.

- People who have lost a loved one or a child feel very isolated. Don't underestimate the power of friendship.

- Remember to be compassionate and non-judgmental.
- Suggest a support group. Maybe even offer to come along, as well.

- Be prepared for a response you didn't expect because everyone grieves differently and some may not be comfortable with verbal communication. If this happens, remember to respond compassionately.

- Remember you cannot take their pain away, refrain from being pushy, just do the best you can to support this person.

It is usually safe to say:

- How are you feeling today?
- I don't know what to say, this must be so hard for you.
- I can't imagine what you're going through, but I wanted to let you know that I'm here for you if you would like to talk, I'm not sure what to say but I can be there to listen.
- Recall a positive memory of the deceased.
- Is there anything I can do for you?
- I miss the person who died.
- Would you like to go for a walk?

All the above-mentioned things are examples of things to say or do to help make a person who is grieving feel comfortable enough to reach out to you for help. Of course, everyone has different sensitivities, so, above anything else, be gentle with someone who is grieving.

The following are some suggestions of what not to say:

Some people often use the term "committed suicide." First off, this term is very outdated and extremely insensitive. "Committed suicide" is offensive because it relates back to an era when suicide was considered a crime or a sin. Instead of saying "committed suicide," say "died by suicide." To help make this more applicable for talking about suicide, substitute the word "cancer" for the word "suicide" to see if the sentence still makes sense or if it has a negative connotation. We wouldn't say "committed cancer" or 'successful cancer." We would say cancer death or died of cancer. Don't say you think it was a selfish act of the deceased. While it may appear that those who die of suicide are not taking into consideration the impact their death will have on the survivors, there is much evidence to the contrary. But the mind of a suicidal person is distorted and often holds the belief that he or she will be lessening their burden on loved ones by no longer being around. Other times their pain can be so overwhelming that they become focused on just ending their pain. Depression, addiction, or other mental health conditions often prevents them from generating alternative solutions to their problems. Many people who have survived a very intense suicide crisis have experienced something akin to command hallucinations right before their attempt.

Never say they're in a better place or they're at peace now. Suicide loss survivors don't see it this way, at least in my experience. It only adds to their pain. Don't say the old cliché, "Everything happens for a reason." What reason would be good enough for your child to die by suicide? Don't say "It was God's will" or "I know how you feel," unless you've lost a child to suicide. Avoid asking details about how the person passed. The bereaved will tell you if they want you to know and if they are strong enough. Avoid hollow reassurance such as, "Things will get better," because it is not helpful to hear.

Don't give your summation of why this happened. Even if your thoughts have merit. It is difficult to process this loss while in shock and any additional opinions may not be warranted. Don't remind the survivor that they are lucky to have other children. Their loss is just as enormous, being just the same as if they only had one child. Try to refrain from avoiding the bereaved as they are grieving and need support. Finally, don't try to rush the survivor through this. They will be ready to move forward when they are ready. You cannot rush someone through the healing process. That will never happen as it is not something one can just "get over." Survivors do find a way to function after a while, but the time it takes varies from person to person. Moving forward takes time and only the survivor knows when they are strong enough to start taking baby steps toward this new chapter of their life which has been turned upside down on them. It is all very unfamiliar and twice as scary. Being the same will never be an option, however, eventually one must make something of the time we have left here on earth. Maybe something to make the person who left us be proud.

How Did I Miss That????

Don't wait until it is too late! Life is precious and fleeting. If you or someone you know is in a crisis situation, call or text one of the numbers below:

*National Suicide Prevention Hotline: 1-800-273-8255

*Crisis Text line: 741741

*Text "Home" to connect with a crisis counselor for help.

<u>Crisistextline.org</u>

www.ingramcontent.com/pod-product-compliance
Lightning Source LLC
Chambersburg PA
CBHW040107120526
44589CB00039B/2791